HIPPOS

For Arusha

Text and photographs copyright © 2005 by Glenn Feldhake

Maps on page 58 courtesy of the IUCN Hippo Specialist Group

Printed in China

05 06 07 08 09 5 4 3 2 1

Library of Congress Cataloging-in-Publication Data

Feldhake, Glenn, 1970–
Hippos : natural history & conservation / Glenn Feldhake.
p. cm. — (Worldlife library)
Includes bibliographical references and index.
ISBN-13: 978-0-89658-726-7 (pbk. : alk. paper)
ISBN-10: 0-89658-726-6 (pbk. : alk. paper)
1. Hippopotamidae. I. Title. II. World life library
QL737.U57F45 2005
599.63'5—dc22
2005011709

Distributed in Canada by Raincoast Books, 9050 Shaughnessy Street, Vancouver, B.C. V6P 6E5

Published by Voyageur Press, Inc.
123 North Second Street, P.O. Box 338, Stillwater, MN 55082 U.S.A.
651-430-2210, fax 651-430-2211
books@voyageurpress.com   www.voyageurpress.com

Educators, fundraisers, premium and gift buyers, publicists, and marketing managers:
Looking for creative products and new sales ideas? Voyageur Press books are available
at special discounts when purchased in quantities, and special editions can be created to
your specifications. For details contact the marketing department at 800-888-9653.

HIPPOS

Glenn Feldhake

WORLDLIFE
LIBRARY

Voyageur Press

Despite their conspicuous size and huge environmental impacts, hippos are perhaps the least researched of the large mammals. Nevertheless, as discoveries are made, hippos never cease to demonstrate unique and surprising physical and behavioral adaptations to their environment.

Contents

Introduction

From 1958 to 1963, the Ugandan Government sponsored the culling of over approximately 7000 hippos residing within the Queen Elizabeth National Park. Ecologists documented the condition and utilization of the habitat both before and after the hunts to determine the role of hippos on the ecosystems. During the hunts, biologists collected a plethora of anatomical data including information on reproductive activity, stomach contents, and the effects of aging. This data collection effort was the most comprehensive to ever occur in the study of the common hippopotamus.

In little more than a decade, the hippo population in the Queen Elizabeth National Park almost completely rebounded. With so many hippos again available for study, German researcher, Dr Hans Klingel, entered the same parks and emerged 12 years later as one of the world's foremost hippo authorities. To date, his studies and observations are the most complete, long-term record of hippo behavior.

While the data collected by the biologists and ecologists in the culling programs and the observations of Dr Klingel provided invaluable data on the hippopotamus, textbooks have often viewed this 'Standard Ugandan Hippo' as representative of the population as a whole. Hippos live in the wild in over 29 countries in Africa. They inhabit a multitude of habitats and ecosystems. Yet the majority of scientific knowledge is based on a colossal amount of data taken from just a few parks in Africa.

As I have traveled throughout Africa to study hippos in the wild and within Europe and North America to meet with top hippo researchers and those who care for hippos in captivity, one point has become abundantly clear – one size does not fit all. With every new African preserve I enter, I hear the same comments:

'We're so glad you're here. Maybe you can answer this for us. All the textbooks say hippos are supposed to do "this" or "that", but the ones around here don't do "that" at all. Why do you think that is?'

The answer is always the same…'We're not in Uganda.'

Hippos maintain a lifestyle that is not necessarily conducive to human observation, but that does not mean there is nothing worth observing. In the last 10 years,

researchers have become increasingly aware that there is a great deal to see, hear, and explain. The existence of underwater vocalizations, a possible common heritage with whales, and the immense influence hippos have on their ecosystems are just a few of the recent discoveries spurring the scientific community to take a second look at the species.

Hippos live in ecosystems with varying predator to prey ratios, seasonal changes, soil chemistry, vegetation, and landscapes. As more is learned across these diverse habitats, hippos never cease to surprise researchers by displaying a multitude of behaviors and adaptations that show the Ugandan stereotypes represent only a sliver of what is to be known about the species as a whole.

Only with a new sense of interest and respect will we begin to learn about the many mysteries and yet-to-be-discovered surprises that this wonderful species is hiding inside its world and its many important roles in the environment.

Origins

To understand the common hippopotamus (*Hippopotamus amphibius*) seen today in waterways across Africa, it is important to first recognize their natural origins and the millions of years of environmental influences that affected their evolution. The hippo's appearance is unique to say the least but is a result of millions of years of evolutionary refinements. Likewise, as ice ages have come and gone and continents have shifted, the ecological niche that the early hippopotamus predecessors have filled has been an ever-moving target.

Recent DNA analyses conclude that the closest living relatives to hippos are cetaceans (i.e., whales and dolphins). This common heritage extends back to a very unhippo-like creature, Pakicetus, which lived roughly 47 million years ago. Without a PhD in paleontology, one would hardly recognize this hippo predecessor as an ancestor of what some of its descendants have evolved into today. Pakicetus was about the size of a bull seal and had webbed feet. These animals used their powerful hind legs and tail to move through the water but could possibly move on land using the same adaptations as modern-day sea lions.

Nothing outwardly resembling a modern hippo arrived on the scene for approximately another 25 million years. The first remains of a species that could be vaguely recognized as a hippo were found in East Africa and date from the lower Miocene, roughly 20 million years ago, though the fossil record is unclear as to how they became more hippo-like as they reached this new part of the globe. These new species were part of a common genus and called Hexaprotodonts, but they still exhibited a number of differences from the common hippo of today. In fact, the Hexaprotodonts bore a much more striking similarity to the pygmy hippo, *Hexaprotodon liberiensis*, of modern-day West Africa. Though many forms of Hexaprotodonts roamed through Africa, most were forest-dwelling animals,

Yawns and gapes are just some of the fascinating behaviors modern hippos display.

spending the majority of their days in thick foliage searching for food while always on the lookout for predators. The Hexaprotodonts were decidedly more terrestrial than modern-day common hippos, with proportionally longer legs more suited for quick escapes and the agility to reach food on irregular terrain. These animals spread across the African continent, evolving into a variety of unique species suited to live in an assortment of environments. In some cases, the Hexaprotodonts are believed to have out-competed other species, such as the pig-like Anthracotheres, for food, leading to their eventual extinction.

Then, roughly 7 to 10 million years ago, a new influence began moving some Hexaprotodonts out of the forests and into more aquatic environments. The impetus for this move is still unknown. Perhaps future research will discover that it was to search more effectively for food, as an escape from predation, or for some other environmental influence. Most likely, the move was a result of some mix of all these pressures. Regardless, the radical change of habitat precipitated a radical change in their anatomy and physiology.

The Hexaprotodonts' new practices of conducting life in an ever more aquatic environment posed many challenges not experienced in the forest. The previous benefits of agility proved less useful than those of pure bulk. In the same way that a small human wading through waist-deep water will be less mobile than a larger individual in the same environment, these earliest of pre-hippos thrived with increasing size. As millions of years passed, they became increasingly aquatic and correspondingly larger.

With added mass, other changes began to occur. In particular, the size of their heads in proportion to their body grew immensely. The distance between their canines expanded and their jaws developed adaptations to allow their mouths to open ever wider. While these many changes may sound strange, if not completely random, they have in fact provided tremendous insights into the lifestyle of hippos throughout their evolution.

When disputes arose over status in the social hierarchy, mating rights, or territorial dominance, the only part of the animal's body above water, and therefore the only part

Even when in the water, any exposed part of a hippo may be visited by birds. Oxpeckers, in particular, frequently cling to the hippos' skin in search of ticks, leeches, and other parasites. This pecking, while sometimes annoying to the hippos, is also credited with helping keep wounds clean.

available to settle matters, was its head. With this obvious restriction, early hippo species began resolving conflicts in matches of weight and strength through 'mouth-to-mouth' combat. The size, number, and position of the incisors and canines all began to shift. In ritualized battles that are still seen on African waterways, the objective for each animal was to swing its head and either catch the opponent's jaw or inflict major gashes with its canines. Once locked, the mêlée became a fierce exercise in pushing and shoving. Such events caused tremendous wear and tear on their teeth and provided the evolutionary pressure to make them ever larger and thicker. Because the acts of pushing and shoving were greatly assisted by extra leverage, the width and breadth of their mouths also expanded continually. Likewise, the position of their eyes, ears, and nostrils also began shifting to the highest part of their skulls, creating a profile similar to that of crocodiles and frogs. With this adaptation, they could remain low in their aquatic habitat while still monitoring the sights, sounds, and smells of the environment around them.

It is important to note, however, that the evolution of hippos took many twists and turns along the way. As long-term climate changes occurred, glaciers shifted, ocean levels rose and fell, and early hippo species were constantly evolving adaptations as a result of their changing environment. Over millions of years, many species of hippos of varying sizes and shapes evolved and eventually met with extinction. Meanwhile, as portions of the world suitable to the various forms of prehistoric hippos came and went, the locations they inhabited were also ever expanding, contracting, and shifting. Three million years ago, hippos arrived in Burma, and as recently as two million years ago, hippos established a long-term presence in India.

During the Cromerian interglacial period, which ended approximately 270,000 years ago, the Earth warmed to a point that hippos began roaming very far to the north. Remains of prehistoric hippos from this time can be found as far north as England. However, these northern-ranging hippos perished with the arrival of the next ice age. Between 120,000 and 80,0000 years ago, the Earth warmed yet again and hippos returned to England via France, bringing with them other tropical mammals including hyenas and narrow-nosed rhinoceroses. Today, one of the most hippo-rich fossil beds in the world is a 100,000-year-old site at Ipswich, England – providing the common geological name 'Ipswichian interglacial period.'

On a few occasions, hippos even inhabited islands hundreds of kilometers from the mainland. At least three species of hippos once thrived on the island of Madagascar at one time or another, but their extinction on that island is now correlated with the arrival of humans 1500 to 2000 years ago. Similarly, the remains of prehistoric hippos on the island of Cyprus are raising many eyebrows among archeologists. Burned and buried hippo bones found on Cyprus have led many archeologists to hypothesize that human inhabitants once considered hippos as a source of food. However, if the alleged age of the bones is correct, it would extend the known existence of humans on that island back at least 1000 years earlier than any other archeological records.

Today *Hippopotamus amphibius* lives throughout sub-Saharan Africa. It is second only to African and Asian elephants as the largest mammal on land. Regional variations occur, however the common hippopotamus is typically recorded to weigh in at more than 4400 lb (2000 kg). Obviously, having acquired such an enormous size as well as a pair of 12 inch (30 cm) canines, a common hippopotamus is in little danger of predation. A lion may only measure 1/10th the mass of a hippo. But, remarkably, the hippo's temperament seems little changed from its prehistoric days in the forests. They remain skittish creatures, easily alarmed, and quick to become aggressively defensive.

Hippos are gregarious creatures, gathering to spend all but feeding and occasional sunbathing sessions together in the water. In fact, with each new discovery made about the hippo's anatomy and even its behavior, scientists become increasingly aware that, despite having legs rather than flippers, the common hippo is not so much a terrestrial mammal that chooses to spend time in the water. Rather, the hippo of today has evolved into a more aquatic mammal that periodically ventures on to land.

The Hierarchy

Hippos are gregarious animals by nature, typically appearing in herds of 10 to 20 animals at a time, though more than 100 are sometimes recorded living together during periods of drought. Herds of these multi-ton animals are conspicuous not only by their size, but by the chorus of unique honks and snorts erupting from the group, which may be heard for several kilometers. Nevertheless, their social habits are not well understood. In large part, this is because for such a remarkably identifiable mammal, hippos are notoriously difficult to observe in any meaningful detail.

They may spend as much as 18 to 20 hours of their day in water that is often filled with silt and strewn with reeds or boulders. Only eyes, ears, nostrils and an occasional backside appear above the surface before submerging and reappearing elsewhere minutes later. What social interactions are recorded are usually taken out of context, as hippo behaviors largely occur underwater and out of sight to observers on the shore. The minimal amount of time hippos spend on land is either used sleeping, offering little to observe, or out feeding in the dark of night when following the animals becomes perilous. In fact, between the diversity of documented behaviors and the difficulty of placing them in any social or situational context, we are fortunate to understand much about hippos. Only through a 'bits and pieces' approach to collecting data about hippo behavior has a bigger picture been obtained.

To date, the single most extensive record of hippo social behavior has come from observations recorded by German researcher, Dr Hans Klingel, a professor from the Zoologisches Institut Universitaet Braunschweig in Germany. Dr Klingel's research found that groups of young males, females with calves, and mixes of ages and genders living in herds under a dominant bull are among the most common social arrangements. However, in the social order of a hippo herd, research indicates that almost anything goes.

Interpreting behaviors requires knowledge of the hippos' age, gender and social status.

Contemporaries of Dr Klingel's, including game wardens, members of academia, and even safari operators, have used his work as a basis to compare their own observations of herds around the African continent. Each community has added to the wealth of individual behavioral observations, yet each has remained stifled by the hippos' aquatic lifestyle. Hippos will make themselves at home in almost any slow-moving body of fresh water. Even the occasional occupation of drainage canals, resort swimming pools, and golf course water hazards has been recorded. But, given the option, a herd of hippos will generally choose rivers or lakes with shallow water and firm, gently sloping beaches, allowing them to sleep in the water without having to lift their heads to breathe.

Hippos are often found lying in physical contact with one another. However, they have also been described as social schizophrenics because at night, when hippos move on land to feed, they seem aloof, almost in avoidance of each other. Nevertheless, hippo interactions are not always civil. Territorial battles, the arrival of a new hippo, or a female going into estrus can lead to a commotion within the herd. And, the interactions between hippos appear to have as much to do with the social rank of each animal as they do with the context in which they occur.

Dominant Bulls

Without question, the most formidable and recognizable member of any herd is the dominant bull. Even as the confusing assortment of eyes and ears pop up above the surface to check on the surroundings, the presence of the bull is usually obvious to a casual observer from the shore. The dominant bull is the largest member of the herd, often moving among and around those in his territory. The bull will periodically produce a loud if not flamboyant vocalization, often described as a 'wheeze honk,' which is in turn responded to in an equally cacophonous manner by the rest of the herd and often by nearby herds elsewhere in the river or lake.

The dominant bull controls his territory along the river or lake shore, but only about 10 percent of male hippos ever become permanent territory holders. For those that do control a territory, unimpeded mating rights are the main trophy, but maintaining 'dominant bull' status can be daunting. The dominant bull is always watching for possible challengers and regularly 'makes the rounds' eliciting subordinate behaviors from others in the herd to reconfirm his dominant status. Territories may constitute as little as 54 to 108 yd (50 to 100 m) of shoreline, but if the habitat remains suitable for hippos to live, dominant bulls have been recorded remaining in control of their territory for more than a decade – a record reign in mammalian circles.

When neighboring bulls meet at their territorial borders, they will typically perform an almost ritualized greeting by first staring nose-to-nose with their ears attentively cocked forward. Then,

Brief skirmishes between hippos are a regular occurrence.

they turn rump-to-rump and defecate while rapidly flapping their tails, sending a shower of dung across each other before peaceably returning to their respective herds.

Despite the bulls' recognition of territorial borders, boundaries can shift. During any season of the year, any number of natural phenomena, such as droughts or fires, can quickly disturb the local ecological balance, forcing hippos to relocate. Likewise, at any time a dominant bull may be challenged for territorial rights by new arrivals or neighboring bulls seeking to expand their borders. If a bull attempts to challenge another over territorial dominance, the result is a violent exchange classically adhering to a strict set of rules. In an initial test of size and strength, the bulls will open their mouths as much as 150° and clash jaws in a match of pushing and shoving. If neither bull relents nor is there an obvious winner, the conflict will escalate and turn bloody. The males will begin swinging their heads, slashing each other with their massive lower canines which may be as much as 1 ft (30 cm) long. Such bouts rarely reach this stage; however when they do occur, it can be a fight to the death.

Males without a territory face difficulties.

Seasonal Bulls

For those bulls not old enough or large enough to hold a permanent territory, some opportunities do occasionally present themselves, though typically it is a short-lived experience. With the onset of rains, water levels rise, and many new suitable habitats with attractive shorelines may become widely available. At these times, many male hippos may become dominant over a territory. However, with so many territories available, one small bull's territory may attract few or even no females.

When the waters recede, the small bulls are faced with several less than attractive options. One option is for the small bull to remain on his now less enticing territory. In doing so, the bull will live a solitary life after other members of his seasonal herd have departed in search of more desirable beaches. However, by remaining, he will have staked an early claim for what could be a good territory to attract females when the rains return the following year.

The seasonal bull may also opt to gather with other non-territorial bulls, in less desirable parts of the river or lake unclaimed by larger dominant bulls, in an arrangement sometimes referred to as a bachelor herd. These less desirable areas may be far from food, have rocky or steep beaches, or simply be along waterways with fast currents.

The final option is to make the risky attempt of living around other hippos in a permanent herd but under the careful and often aggressive watch of a dominant bull. In order to stay, the younger or smaller bulls must adhere to the dominant bull's rules. Subordinate males must behave submissively and not try to mate with the females. Still, even when on their best behavior, the small bulls are often subject to frequent aggression from dominant bulls and may eventually choose to move elsewhere.

Females

Groups of females represent the only semi-stable social units containing both young and old hippos. Females with their offspring may travel together, and daughters frequently remain with their mothers while approaching adulthood. Male calves remain with their mothers and the other females in the herd but are ousted around the time they reach sexual maturity – anywhere from age 3 to 15 depending on habitat.

As long as the water levels remain satisfactory and there is enough food available nearby, females seem content to stay in the territory of a dominant bull. However, when environmental factors dictate, females, sometimes in small groups, will gather their calves and leave the territory in search of a better habitat. There is little a dominant bull can do to entice the females to stay.

Females arriving at a new territory are readily welcomed by the dominant bull. However, research indicates that the females' final choice of habitat has more to do with the quality of the habitat than the attributes of the bull controlling it. The females, with their calves, will usually spend their day on the best resting places on the beaches for sunbathing, nursing, and sleeping. Outwardly, such privileges appear to be an incentive provided by the dominant bull in order to encourage the females to stay. However, some observations

Regardless of status in the herd, much of a hippo's day may be
spent napping and sunbathing on the beach. However, even small disturbances can
cause the entire herd to quickly return to the perceived safety of the river.

indicate this condition is mandated by the females. The females outnumber the lone bull and may attack in groups from the side and rear rather than observing the ritualized fighting 'rules' the males afford each other.

Maintaining a presence on the best beach locations offers the most ideal conditions for raising young, which may be cared for by multiple females rather than just the mother. At night, as the herd leaves the water to feed on land, mothers may leave their calves in the care of other females. Such cooperation also extends to the feeding of young calves. In field studies, the number of lactating females is often reported being two to three times the number of calves of the age to be nursing.

Calves

Hippo calves are often found congregated together, spending their time much as many young mammals do. They interact socially but learn from those interactions at the same time. By roughhousing and playing, calves learn the social order of their herd and the basic skills needed in order to survive once they have reached maturity. They will often take part in bouts of mock fighting and someday use the skills they have learned to either defend their young or battle for a territory.

Hippo calves, like the youngsters of so many species, also learn from their natural curiosity of their surroundings. In this regard, hippo calves seem to have an inexplicable fascination with dung, in particular that of the dominant bull. In the wild, hippo calves will sometimes lick and smell around the anal regions of the dominant bull, or even follow dominant bulls onto land when they defecate. Often hippo calves, sometimes accompanied by adult females, will explore the beaches, sniffing and sometimes nibbling at the dung scattered along the banks. Theories for this fascination abound yet all are speculative. They include the calves' ability to detect chemicals in the urine identifying age and gender, as well as the calves' interest in understanding to whom they are related.

Evidence does not necessarily show that bulls are equally fascinated with calves. In fact, any undue interest shown by a bull toward a calf could mean danger. By some estimates, the number one killer of young hippos is other hippos, particularly bulls. Female cooperation within a herd may even occur, in part, as a protective measure against the threat bulls pose. The killing of infant animals by adults, 'infanticide,' is not uncommon or unique to hippos. It has been recorded in primates, birds, and rodents. One of the more plausible explanations for this behavior in hippos is that bulls are simply trying to improve their reproductive success.

Females with calves typically go into estrus twice while they are still nursing their young. The first occurrence happens approximately 50 days after the birth of their calf. However, nursing a calf imposes a large physiological stress on the female's body, causing her to be less likely to conceive. If a bull kills the young calf, the stress of nursing is lifted and conception will be more likely to be successful.

Corroborating observations indicate that the ages of the calves killed by bulls tend to be between one and 40 days old. Further, infanticide appears to occur most often during the dry season. This is when the number of available territories diminishes, males are most likely to be ousted from their dominant positions, and females with their calves arrive in new habitats. Such observations lend support to the idea that male hippos may kill calves fathered by other males in order to gain sexual access to the local females and promote their own offspring.

If not killed by the bull in the first couple of months, there are still many dangers facing a young hippo. In its first year of life, anywhere from 15 to 40 percent of calves may die. Mortality rates are roughly halved in the second year and become negligible by year three. Most often, deaths not due to bulls may be attributed to either predation or complications from malnutrition during periods of environmental stress. But, in the wild, regardless of age, each day poses its own challenges for survival to all the herd's members.

A Day in the Life

For hippos, a typical day does not conform to a human timetable. The business of the day actually begins around early evening when the sun gets low and the air begins to cool. At this time, the hippos begin to wake up with a yawn and a stretch to prepare for an evening of feeding. Cacophonous snorts and honks pass through the air with increasing regularity. As more hippos wake up, they disturb those still snoozing in the herd, causing swipes and brief skirmishes. Soon, all are awake and ready to begin their night of feeding.

Off to Work

When the sun finally sets, it is time to feed. Feeding takes place largely in the dark of night and begins as the hippos venture from the water via 'gateways' connecting their aquatic life to their terrestrial wanderings. These gateways are developed over time as the hippos exhibit the seemingly obsessive behavior of always entering and leaving the waterways at exactly the same point. Over weeks, months, and even years, the repeated trampling of the ground by these multi-ton animals, along with the natural erosion created by water lapping along the beaches, causes ruts to form that are

A well-worn hippo 'gateway'.

sometimes over 6 ft (2 m) deep and exactly one hippo wide. The sides of these gateways may become polished smooth by the sides of the hippos rubbing against the exposed earth as they move single-file off to their grazing areas.

Along the way, hippos are no less habitual about their choice of route. In fact, hippos have developed a reputation in Africa as the great trail-makers of the bush. These trails are followed so religiously by the hippos that, over time, they become uniquely identifiable by their double-grooved structure. The right pair of feet walk in one groove; the left pair of feet remain in the other. These trails are kept open through even the densest brush by the hippos' repeated wanderings and are used during the day by many other species of all shapes and sizes seeking a clear path between the grasslands and the water.

As the hippos traverse their trails, they will frequently stop to add to prominent dung piles located on trees, boulders, or even bends along the way. These dung piles are usually most evident close to the water, and their significance has generated considerable speculation. Early research assumed them to be a form of territorial marking. However, more recent observations have shown that hippos of varying ages and genders will add to the sites, even hippos from other herds. Such behavior would cast doubt on a strict 'territorial marking' explanation. Today, new theories suggest that the dung piles are placed on landmarks along the trails, and the scent serves to help the hippos find their way between the grasslands and waterways in even the darkest of nights (i.e., navigation by nose). Nevertheless, all explanations put forward thus far remain somewhat speculative.

When the hippos arrive at their grassy feeding areas, each ventures off in its own direction in search of short, soft grass. Based on studies of stomach contents of hippos killed in Uganda during the late 1950s and early 1960s, hippos are widely reported to eat nothing else. However, observations have been made elsewhere in Africa that indicate hippos will occasionally eat aquatic vegetation, fallen fruit, or even nibble on bushes.

Hippos will maintain their grazing areas with the same constancy as their trails and gateways. These areas, sometimes referred to as 'hippo lawns', may only be 54 yd (50 m) wide, but may extend for more than 0.6 miles (1 km) and have the appearance of a well-trampled golf course fairway. Though hippos could quickly fill their stomachs with tall grasses growing along the edges of their lawn, they seem content to consume only the short, tender grasses they work so hard to maintain.

By morning, an adult hippo has consumed approximately 110 lb (50 kg) of grass – a relatively small quantity of food when compared to other species. However, the combination of a very efficient digestive system absorbing a high amount of nutrition even from low-quality food, and the low energy requirement of remaining in a semi-buoyed state in the water most of the day, reduces a hippo's need to eat large quantities. Other large mammals, including elephants, rhinos, and even African cattle, will consume around 2 to 2.5 percent of their body weight in food each day. As a percentage of body weight, a hippo will only eat about half as much.

Back to the Beach

As the sun begins to rise, the hippos return down the trails, through their gateways, and splash back into the rivers and lakes along their beachfront territories. Honks and snorts echo through the bush with the arrival of each member of the

A network of hippo pathways through the Okavango Delta.

herd. The hippos may nap in the water after their meal, but many of the social aspects of hippo life often take place at this time. The greatest commotion can occur with the arrival of a new hippo to the beach. Hippos are not prone to move to a new area without reason. However, those that arrive at a new habitat may have traveled several kilometers. Females and their calves will typically only move to new territories during the changing of seasons, when either the water level or the availability of nearby food become unacceptable. Young male hippos, however, may move around a bit more frequently. Young males will attempt to

Honks and snorts will erupt from the herd as each hippo splashes back into the water after returning from a night of feeding. Hippos regularly return to the same beach each morning as long as the environmental conditions remain favorable.

join herds living in favorable habitats but, as they are also subject to frequent aggression by the dominant bulls controlling the territory, they may soon decide to leave.

The dominant bull will be quick to 'greet' any new arrival attempting to join his herd. If a new male arrives, the dominant bull will aggressively ensure the new male recognizes his status, and will make his status known with an assortment of behaviors. Staring intently at the new male and keeping his ears attentively cocked forward, the bull may arch his back high out of the water to demonstrate his massive size as a sign of intimidation. The bull may also lift his head high out of the water, opening his jaws a respect-provoking 150° and showing off his 1 ft (30 cm) canine teeth, or defecate while rapidly flapping his tail in a display sending dung showering in every direction.

If the new male is to stay, he must respond with a sign of submission to the bull's authority. Most often the new male will raise his rump high out of the water, dropping his feces without the flamboyant display of the bull. Or, he may keep his body low to try to look too small to be a threat. Regardless of how convincing the new male's demonstration of submissiveness may be, the newcomer may be attacked and chased away from the herd many times before being allowed to remain. The dominant bull will spare no effort to maintain his position, and any male perceived as a potential threat to his control of the territory will not be allowed to stay.

At other times, females, sometimes with calves, will arrive, leading to a very different response from the dominant bull. The bull will readily allow the females to settle in to his territory and his attention will almost immediately focus on determining if a female is in estrus. The bull will regularly pass from female to female in his herd, testing their urine with his vomeronasal (or Jacobson's) organ. The Jacobson's organ appears as a pair of openings located in the roof of the hippo's mouth. Most mammals have a Jacobson's organ; humans are one of the few that do not. The organ functions differently from species to species but is used to detect pheromones and/or hormones of members of the opposite sex that identify reproductive readiness. In hippos, the Jacobson's organ is considered huge, though little is known about what chemicals the hippos are specifically trying to detect.

Mating, like almost all aspects of hippo behavior, takes place in the water. When the bull determines a female is in estrus, he pursues her relentlessly. In many academic references, breeding has been described as a violent affair, initially confused by one scholar as two animals locked in 'mortal combat.' However, others have documented and recorded a subdued event that can easily go unrecognized by a casual observer.

The act of mating can last up to a half an hour. During copulation, the female lies down in a submissive position as the male climbs on her back from behind. Frequently, the female's head is forced underwater. After mating, the female will return to the other females in the herd, and, if the mating was successful, her pregnancy will last eight months.

In the wild, the time between births is short. Hippos are regularly estimated to give birth every two to three years. One study in Uganda showed, over the course of 18 months, 15 to 32 percent of all female hippos were pregnant at any given time. This number may vary from habitat to habitat around Africa as nutritional and social variables shift. However, the figures from Uganda are likely consistent with much of the continent.

As the delivery time approaches for her calf, the mother will leave the herd to give birth in shallow water or occasionally on land. According to one report, there are specific places regularly visited by females to give birth, but no other independent observations of this habit have been made. For the first few days of life, the mother constantly attends to her calf, forgoing even her nightly feeding. The pair will remain separate from the herd for an 'imprinting period,' which lasts approximately two weeks. The primary purpose of this time is for the calf and the mother to learn to identify each other. Hippo calves appear to have a vague concept of 'mother.' If the imprinting is not successful, the baby may go chasing after any large moving object – even a car. During this phase, the mother will drive away all others, including her own offspring from previous years. The exact mechanism by which the calves learn to recognize their mothers is as yet unknown. However, only when the two begin to identify each other with certainty will they return to the herd.

As the Sun Gets High

As the day warms up, activity appears to settle in the herd and few interactions are seen from the shore. Periodically, the bull may raise his nostrils above the surface and unleash a high-pitched wheeze, immediately followed by several short but powerful staccato honking sounds reverberating across the water. In response, noses begin appearing around the herd to also chime in with similar honks and snorts. In the distance, other herds may be heard chiming in one-by-one sequentially down the waterway.

The purpose of these abrupt interruptions to the serene setting of hippos quietly resting in the water was considered little more than a curiosity until the late 1980s, when an amazing and unexpected discovery was made. In March 1989, Dr William Barklow, a biology professor from Massachusetts in the United States, went on a personal vacation to Tanzania and made a few off-hand observations that led him to believe hippos were actually communicating vocally underwater – sounds not heard above the surface. Two years later, he returned to Africa, armed with underwater recording equipment, and proved it.

Whereas hippos were once thought to make little more than those few ear-splitting honks and snorts through the air, we now know they have a varied and complex vocalization system that is used underwater and out of earshot of humans. The calls sound, in the most basic sense, something like those of their distant cousins the whales and dolphins, but are still unique to hippos. The underwater calls are difficult to separate into precise categories, but roughly about a dozen different types of clicks, croaks, and whines can be identified. Many of these calls are made exclusively under water; however, hippos can also create and listen to sound above and below the surface simultaneously. This ability lies with the hippo's evolution of a second simple but separate underwater acoustical system which roughly parallels that of dolphins.

This mechanism allows hippos to overcome pitfalls experienced by most mammals when their eardrums are submerged. In most species the bones of the middle and inner ear are deeply embedded in the skull for protection. When submerged, sounds passing through the water reach the eardrum by vibrating through an animal's entire skeletal structure, making the sound muffled and seeming to come from all directions. In hippos, as well as porpoises, however, the middle ear and the inner ear are partially suspended from the skull by ligaments in a way that would reduce random bone conduction.

In hippos, the middle ear is held in place by three attachments. One connects the middle ear to the outer ear; the second connects with the cochlea. The third attachment, however, is curiously connected to the portion of the skull that makes contact with the jaw. Similar to a structure found in porpoises, each side of a hippo's jaw contains a remarkably thin portion of bone which is well suited for acquiring sound as it passes through the water. With this adaptation, a hippo may monitor both its terrestrial and aquatic environments while its jaw is under water but its external ears remain above the surface.

Similarly, the way in which hippos create sounds in both environments may also be attributed to a mechanism paralleling that of dolphins – by vibrating fatty layers. When a hippo vocalizes through the air, the sound is produced in the larynx. However, the call is often emitted through its nostrils as the jaw and throat remain below the surface. Hippos have a large roll of fat located across the larynx just under the throat. If hypotheses are correct, as the hippo's larynx creates sound through the air, vibrations in the roll of fat propagate the sound simultaneously through the water.

The meanings of their calls are still largely unknown, but several intriguing hypotheses have been suggested regarding the complexity of the information hippos may obtain. For example, as sound travels at different speeds through air and water, it is possible hippos can

*Hippos spend much of their day napping together in the water and on the beach,
often in physical contact with each other. Any jostling by a member of the herd changing position
or moving to a new spot will ripple through the group, causing swipes and brief skirmishes.*

determine from the delay (or apparent echo) the distance to the next herd. Females may be able to use this information to determine where and how far they will have to go in the event they need to leave the territory they are living in. Bulls could also possibly use this information to monitor neighboring bulls.

Dr Barklow later stated of his findings, 'About 80 percent of their vocalizations are given underwater, and the underwater sounds are the ones we know the least about. They probably indicate a level of social organization no one has ever suspected.'

Time to Sleep

When the temperature has peaked it is time for the hippos to get some sleep in preparation for another night of feeding. Hippos will sleep often in direct contact with each other, lined up on a beach or in the water with each one's head resting on the rump of the next, looking something like a chain of fallen dominos. To the casual observer, watching a herd of sleeping hippos may appear to be one of the least interesting activities imaginable. However, in the last several decades, scientists began noticing that even as the hippos slept, they were demonstrating behaviors that raised many questions and led to a few surprising answers.

Remarkably, though hippos are almost completely hairless, they can often be found asleep on beaches, lying in the blazing sun rather than taking refuge in the water. This behavior has even been documented in the few cases of albino hippos – all without any outwardly ill effects of overheating or sunburn. In search of the reason for hippos' tolerance to the sun, researchers discovered a rather unique substance that hippos themselves create. Hippo skin exudes a bright red substance, often referred to simply as 'hippo sweat.' The color of this fluid is so striking, at one time it was believed that hippos were sweating blood. The substance, though, is neither sweat nor blood. It is created by glands in the skin and is actually opaque to ultraviolet light. In short, a hippo's skin generates its own natural sunscreen.

Perhaps even more remarkable, hippo sweat has since been shown to offer medicinal value to wounds from fights and scrapes suffered from the vegetation along the edges of their trails. The water in which hippos reside is usually warm and filled with algae and bacteria. However, despite what would intuitively seem like a perfect breeding ground for infections, the incidence of infections is rare due to a chemical found in hippo sweat called 'hipposudoric acid.' When added to forms of pathogenic bacteria in laboratory tests, the bacteria stopped growing. The glands of the hippo's skin appear to generate their own shield of antibiotics.

Those hippos preferring to sleep in the water can do so, remaining submerged for about five minutes at a time. And, in the same way humans might roll over in the night without waking up, the hippo will raise its nose just above the surface to catch a breath and then go back down, all while remaining completely asleep. The mechanism for how hippos can hold their breath for so long, however, was for many years another source of academic debate. Early theories suggested that hippo blood must contain a high amount of the oxygen-carrying substance, hemoglobin. However, subsequent tests concluded a hippo's blood retains oxygen about as efficiently as that of most other mammals.

In reality, as a hippo holds its breath, the build-up of carbon dioxide in its bloodstream causes its heart rate to drop – a phenomenon called bradycardia. Blood flow is reduced to parts of the body not being used, but remains constant where it is still required, creating a more efficient distribution of the oxygen. The heart rate of a hippo walking on the river bottom may be only half of what it is when breathing air, and when asleep, the hippo's heart rate may drop to as little as a quarter of normal. This phenomenon is not unique to hippos. It is known to occur in porpoises and even some species of birds which dive for fish. However, it is yet another example of how hippos have evolved many adaptations to survive and play an important role in the habitats in which they live.

Playing a Role

There can be little doubt that groups of multi-ton animals are destined to have a tremendous impact on their environment. In the case of hippos, opinions of their positive or negative effect on the environment often appear to have more than a little subjective influence. And, among ecologists and conservationists, even the most basic assumptions may be of considerable dispute.

Grasslands

The majority of debates have centered on hippos' specific manner of grazing and the associated effects of the creation and use of their trails and gateways. As a hippo feeds, it grabs bunches of grass with its wide lips and swings its head to the side, tearing off the grasses just above the surface of the soil. This feeding style keeps the grasses short and maintains the manicured appearance of a lawn. The tearing, however, also loosens the root structure holding the topsoil in place. With the onset of rains, the deeply worn hippo trails connecting the lawns with the waterways serve as efficient drainage channels, carrying the loosened topsoil directly into the rivers.

In some areas, more than 80 hippos have been recorded feeding within a 1 sq. mile (2.6 sq. km) grazing area. Under such conditions, the native grasses may be completely stripped away. Such an overburdening of a particular habitat by a single species often raises fears that the diversity of plants and animals required to maintain the local ecological balance will not be preserved, and has been used as the justification for many hippo-culling programs. One of the largest and most well-documented hippo-culling programs was initiated in the Queen Elizabeth National Park, Uganda, for just this reason. The culling program began in 1958 and by 1963 had eliminated approximately 7000 hippos. Researchers took advantage of the opportunity provided by the scale of the program, documenting and measuring the overall environmental health of the park both before and after the cull. Among their more striking findings was that in areas where hippos had been completely removed, there were dramatic increases in the amount of vegetation within the first mile of the waterways. Elsewhere in the park, the diversity of both animals and vegetation was also seen to increase greatly following the killing of the hippos in those areas. In some places, as many as 208 species of mammals began living where only 89 had previously existed. With such an increase in biodiversity, parks officials declared that the projected environmental disasters had been averted and the program was a resounding success.

However, individual success stories do not necessarily justify the automatic use of hippo culling to resolve all future overpopulation problems. In fact, even determining when a population has grown to the point of being a 'problem' has proven itself to be very difficult. At the same time as Ugandan park officials were trying to make drastic reductions in hippo populations, at least as many hippos were living in the Virunga National Park in neighboring Democratic Republic of Congo (formerly Zaïre). In the Virunga National Park, ecologists believed that the hippo population of 19,000 was simply re-establishing itself to a natural level not seen since the arrival of colonial hunters in the early 1900s. A strict 'hands-off' policy was instituted. No shootings occurred, and a thriving ecological balance eventually established itself – albeit a balance with an enormous number of hippos. Fears such as those in Uganda that an overpopulation of hippos would lead to the permanent destruction of the habitat for all species were not the result in Virunga. Outcomes such as these have since led some to call into question if the Ugandan hippo-culling programs were ever justified.

Today, some conservationists are identifying circumstances where the role of hippos is even beneficial to grasslands. In more recent research from the Queen Elizabeth National Park, Uganda, some hippo lawns have been found to play a vital role in preserving that ecosystem's overall health, although this conclusion came after the damage was done. In some areas where hippos were removed to protect the grasses, the benefit was short lived as the loss of trees and shrubs during subsequent brush fires increased dramatically. In short, the mosaic of hippo lawns crisscrossing through the bush can serve as natural fire lines. When hippos are no longer present to maintain the lawns, fires spread unchecked. Further, in Kenya, a balance of fires and hippo lawns has the beneficial effect of rotating local vegetation. After a fire has removed an area of trees and shrubs, the burned area begins the cycle of growing new grasses which attract hippos to feed. Reciprocally, old lawns which the hippos have stripped of grass are soon abandoned, allowing the growth of new trees and brush. Such a rotation of vegetation also shifts the wildlife that are dependent on it for food, as well as their predators.

These observations are important as they reflect a significant shift in the views of ecologists over the last several decades. Whereas the concept of an ecological 'balance' has historically been viewed as static, the importance of change is now increasingly recognized. And, in this respect, hippos appear to provide their greatest benefits. Researchers documenting the dynamics of Botswana's Okavango Delta note:

'During the course of 10 years of research … we have noted numerous instances where the role of hippos appears to have been pivotal in determining the ecosystem's response to changing conditions, and … we compile these observations to emphasize the importance of the hippopotamus in this remarkable ecosystem.'

Waterways

The ecological consequences of hippos in African waterways have also been the subject of debate. The areas where hippos feed are not necessarily where they defecate. Each night, a herd of hippos may consume enough grass to provide thousands of pounds of natural fertilizer to the waterways where they spend their days. After re-entering the water following a night of grazing, hippos spread their feces rather flamboyantly, in a practice called 'dung showering.' As a result, they redistribute large amounts of nutrients from one area to another. This behavior has been identified to promote the growth of non-indigenous plants such as the choking aquatic weed *Salvinia molesta*, and the water hyacinth, *Eichhornia crassipes*, in Botswana's Okavango Delta. The spread of these plants has subsequently led some to suggest new hippo-culling programs as the best way to prohibit their spread.

However, the same dung-showering behavior which has drawn condemnation from some has also long been credited with providing benefits to their aquatic habitats. With the spreading of their feces, indigenous alga and aquatic plants are provided with much-needed nutrients to grow. The alga provides food for small fish and the plants offer them a place to hide. The presence of small fish, in turn, attracts larger fish in search of food. The larger fish attract aquatic birds, and soon a healthy ecosystem may develop. In Lake Victoria, Uganda, an estimated one ton of fish was once estimated to exist in each acre of the lake. To many researchers, the presence of hippos was the obvious reason.

The hippos' aquatic habits have also been recognized to benefit the reproduction of certain fish. In fact, in the Ndumu Game Reserve in South Africa, the preservation of entire species has been attributed to the aquatic trails created by hippos by linking seasonal floodplains used for breeding with the Pongolo River. During the flood season, several species of fish, including red-nosed mudfish, purple mudfish and tigerfish, use the hippo trails as 'feeder' channels between the two habitats. Adult fish enter the pans from the river to breed and lay eggs. Meanwhile, fish hatched from the previous year use the flooded channels to enter the river for the first time.

Increased attention to observations such as these is slowly leading to a greater recognition that the value of hippos in their native ecosystems often outweighs the negative consequences of their presence such as the spread of non-indigenous weeds. For this reason, the amounts of research and conservation resources devoted toward hippos have slowly begun to increase. Those working to protect African aquatic ecosystems in general have come to realize their limited conservation resources can be used more efficiently by ensuring hippo populations remain healthy. If the hippo population remains healthy, the odds are raised that the entire ecosystem will as well.

Cooperation

Other species have also been found to benefit through a variety of synergistic as well as parasitic relationships with hippos. At least 18 taxonomic families of birds have been identified to either rest safely on hippo backs, use the rump of a hippo as a perch for hunting small fish and invertebrates disturbed as the hippos walk along the river bottom, or simply feed on ticks or other parasites.

One of these parasites is the little-known *Placobdelloides jaegerskioeldi*, or 'hippo leech.' This creature has made itself noteworthy by evolving many unique, if not bizarre, adaptations to allow itself a parasitic existence solely with hippos. With its flattened body shape, the hippo leech can ensure it is not swept off the hippo by fast-flowing water, and its coloring matches hippo skin, making it well camouflaged from oxpeckers. In experiments, the hippo leech has shown it will attach itself to no species other than the hippopotamus. Such species-specific behavior is undocumented elsewhere in the leech world, making this species a truly unique leech among leeches.

With so many observations of the interdependency between hippos and their ecosystems, as well as with those with whom they share it, their conservation is obviously important to more than just hippos as a species. At worst, the value of hippos to their ecosystems, is still a point of debate. But, evidence continues to offer that millions of years of evolution have molded them to fill a role as a contributing part of a living eco-system.

A Thing of Beauty is a Joy Forever

Hippos have existed on Earth for approximately 20 million years, or roughly about 10 times longer than humans. As of yet, they do not appear in any immediate danger of extinction, but they are certainly subject to many of the same environmental pressures as other African wildlife. Habitat loss, poaching, and environmental encroachment have caused the population of hippos across the continent to decline greatly over the last hundred years.

Today, possibly the most significant commercial pressure leading to the killing of hippos has been an increased consumer interest in their 24 in (60 cm) long canine teeth as a substitute for elephant ivory. The practice of carving hippo teeth is not new. Records of collecting hippo teeth for carvings date to the time of the ancient Egyptians and the trade in hippo ivory has been an established industry in the West for decades. In 1934 approximately 7000 lb (3175 kg) of hippo teeth were exported just from Tanganyika (now mainland Tanzania) to the outside world. The following year, Uganda exported almost 4600 lb (2100 kg).

Demand for hippo teeth for use in carvings has increased.

According to the United States Fish and Wildlife Service, beneath the enamel layer, hippo teeth are the chemical equivalent of elephant ivory. Nevertheless, the demand for hippo ivory has historically always been insignificant compared to that of elephants, until 1989 when the international trade ban on elephant ivory left traditional carvers in search of new, legally exportable materials. In 1988, one year prior to the ban on elephant ivory, the whole of Africa exported around 5600 lb (2500 kg) of hippo teeth to the outside world. By 1991, two years following the ban, exports had increased by about 600 percent to almost 30,000 lb (13,600 kg) per annum and have remained relatively steady to date. Over the same period, smuggling of hippo teeth rose dramatically as well. In May of 1997, customs officials at Orly Airport in Paris confiscated 1738 hippo teeth from smugglers. The teeth were *en route* from Uganda to Hong Kong. In June 2001, 10,000 lb (4500 kg) of hippo teeth were seized from the home of a trader of live animals in Uganda. An estimated 2000 hippos were killed to gather a collection that size.

It is important to note, however, that not all hippo hunting is for commercial gain.

Snares pose threats to all African mammals.

Subsistence poachers, those who illegally hunt for food, are also a concern, though to a lesser degree — largely for practical reasons. Hippos are widely regarded in Africa as the most dangerous animal to humans. The hunting of just about any other species proves considerably less perilous. Further, once killed, the task of moving the meat from a multi-ton carcass to a village is no less trivial.

Those who do poach hippos for food often do so using snares. Hippos are highly susceptible to snaring because their paths are so easily identified. Winch cables from trucks, guide wires from utility poles, and even cable fencing surrounding the preserves where hippos live have been used. The true danger of snares, however, is that they are indiscriminate and will constrict around any animal, including elephants, large enough to pull the loop closed. Still, the attempted hunting of hippos for food, even legally, is not widespread. Many textbooks report that hippos are closely related to pigs. As much of Africa is Muslim, hippos are strictly forbidden as a source of food. And, in Zambia, where a major percentage of the world's hippo population still lives, the eating of hippo meat is falsely believed through tribal lore to cause leprosy.

In response to declines witnessed across the continent, some legal protections have been afforded to hippos via an assortment of domestic regulations and international agreements. At an international level, the primary agreement providing protection to hippos, as well as for most endangered species, is the Convention on International Trade of Endangered Species of Flora and Fauna (CITES). CITES provides protection by means of listing species in one of its appendices. Listing under Appendix I, offers a ban on any international trade (i.e., importing and exporting) of products derived from the species listed, such as elephants and rhinos. Under Appendix II, where hippos are currently listed, the international trade of products obtained from the species may take place, but within strict regulatory controls. Documentation must accompany any products taken across international borders, identifying their origin and showing proof they were taken legally.

Monitoring the effect of such agreements, providing scientific data of conservation status, and even offering grassroots public education often falls on conservation agencies and non-governmental organizations. The World Conservation Union (IUCN), one of the largest and most important in the conservation of hippos, conducted the most recent estimate of the world's total hippo population in 2004. To conduct the census, the IUCN Hippo Specialists Group compiled a combination of questionnaires sent to researchers and representatives of national wildlife agencies as well as surveys of literature from academia, non-profit wildlife organizations, and reports of national wildlife agencies. The results were used to establish a country-by-country assessment of the numbers of hippos, levels of protection, and major threats to the indigenous populations.

The current census estimates that between 125,000 and 148,000 hippos are living in the wild. Hippos are distributed across at least 29 different African countries, of which declining populations are reported in half. Most of the countries with declining populations are scattered through the western nations of sub-Saharan Africa including Benin, Guinea, and Ivory Coast. Hippos appear in abundance in East Africa, but with greatest overall populations throughout Southern Africa. While in many countries the numbers of hippos are considered to be stable, only in Zambia, and possibly Uganda, are the populations thought to be increasing. Unfortunately, in the past decade, the former hippo strongholds of the Democratic Republic of Congo saw population estimates plummet as a casualty of political instability. Simultaneously, Zimbabwe and Burundi are also thought to be experiencing massive declines for similar reasons.

To address these declines, the development of a sound conservation strategy must include generating support at grass roots level. However, in the case of hippo conservation, garnering this support poses its own significant challenges. Across the continent, hippos are not only reported to be one of the most prodigious crop raiders but also the number one killer of people. Much of the reason for the frequently tragic interactions between hippos and humans is that the hippos prefer a habitat with an abundance of fresh water and nearby flat grassy land – the same habitat preferred by humans. During the day, hippos sleeping beneath the surface are often awakened unexpectedly by fishermen passing over them in small dug-out-style boats. At night, when the hippos leave the water to feed, they may be startled by human passers-by. In either case, the skittish nature of the hippo, an apparent evolutionary remnant from millions of years of being a small animal subject to predation, is likely to make them charge or attack.

For this reason, hippos in many countries are often viewed as more valuable dead than alive. To commercial and subsistence farmers, living hippos represent the potential for significant crop losses. In Malawi, for example, frequent falsified claims of damage to rice crops are made by villagers to wildlife officials, in order to have the hippos killed, not only as a pre-emptive measure to avoid the possibility of future damage but also to obtain fresh meat. Additionally, with continuing habitat losses due to the expansion of agriculture, even hippo populations residing within the perceived isolation of the boundaries of wildlife preserves are being squeezed out into areas inhabited by humans and thus in direct conflict with them.

All these pressures result in the simple reality that for hippos, as well as many other species, the concept of 'living in the wild' today is not what it was 50 years ago. Even when life for a hippo occurs without human interaction, it is not without human intervention. The development of wildlife management plans in parks means hippos are subject to culling programs and planned burns of their habitat. Outside of park boundaries, damming of waterways for energy or draining for irrigation will also have measurable effects. When use of publicly protected lands is made available to individuals, or dwindling resources force hippos out of park lands, the consequential challenges of hippo/human cohabitation are not trivial.

In many countries, park and wildlife agencies sponsor hunts to reduce environmental pressures or remove 'problem' animals. The hunts most frequently occur to keep the size of the hippo population within the limits of what the habitat can sustain, but also consequentially

The bones of a hippopotamus may be crushed by hyenas or gnawed on by porcupines.

provide food for local inhabitants and revenues to fund other wildlife management programs. In recent years, park and wildlife officials have found that, rather than paying professional rangers to do the hunting, outsiders with deep pockets and dreams of tracking big game will pay to do it themselves. This has led to an ever-increasing price on the heads of wildlife in Africa. In 1993, the cost of a hippo license in Tanzania was a mere $840 but by 2002, some hippo licenses could fetch as much as $3700. While these revenue streams would outwardly seem beneficial, unfortunately, in many African countries, allegations surface all too often that funds raised are never seen by the conservation programs they were intended to support – in fact the funds seem to disappear altogether.

In South Africa, however, views are changing as the parallel niche industry of exotic animal sales has turned into a big business, with potential benefits to both the wildlife and the human populations, as the programs are implemented with proper transparency and public oversight. Here, rather than killing surplus hippos or those creating public safety risks, the hippos are captured and auctioned to independent parks and ranches, providing revenues for often resource-limited park rangers and wildlife programs.

Meanwhile, hippo conservationists are also profiting – not financially but from continuously improving technology. Perhaps the greatest benefits have come in the form of tools for assessing the basic environmental needs of hippo herds, ascertaining the herds' overall health, and determining the health of the ecosystems where they live. Methods of data gathering available to researchers today are leaps and bounds ahead of those available to researchers in the 1960s, and new techniques are constantly being developed.

Today, geographic information systems (GIS) offer researchers the ability to collect data digitally, store it, and then correlate it with information obtained in other studies, resulting in wildlife conservation plans customized for individual ecosystems. Perhaps one of the most exciting technological developments is the arrival of conservation genetics. With ever more isolated hippo populations and increased knowledge and techniques of investigating DNA, researchers with the IUCN are now embarking on programs to collect genetic material from hippos. This will be used to assess the amount of genetic isolation (i.e., inbreeding) within hippo herds, so that conservation resources may be allocated where they are needed most, more effectively ensuring these wonderful animals may continue to thrive and fulfill their important environmental roles across the waterways of Africa.

In 2005, more than 350 hippos were living in zoological parks around the world, serving as ambassadors for their species by both educating and entertaining visitors. To provide better insights for the public, several new multi-million dollar hippo exhibits have been built over recent years that include underwater viewing.

The Future

Scientific views on even the most fundamental aspects of hippo life are constantly changing. In academic and popular circles alike, hippos have been classically viewed as 'land mammals' that choose to spend the bulk of their time in the water. However, each new bit of modern research paints an ever-increasing picture of hippos as a semi-aquatic (if not truly amphibious) species that spends a bit of its time on land. With this realization, it seems the scientific name, *Hippopotamus amphibius*, originally applied in 1758, may have offered the most accurate description of the species. Regardless, the world is becoming increasingly aware that there is always more to learn about hippos beneath the surface.

As scientific awareness increases, the greatest points of interest for next-generation hippo-related studies will not be to just collect more of the same Uganda-style data over wider areas of the African continent. Rather, researchers will have the challenge of supplementing the earlier studies across wider areas while still answering new questions. To this end, a much wider array of data-gathering techniques will have to be employed.

The technologies available to researchers today surpass those of days gone by and new methods for learning about hippos are being developed all the time. Terms such as 'conservation genetics', 'satellite tracking', and 'geographic information systems', unheard of in the 60s, represent new tools in the collective arsenals of ecologists, biologists, and wildlife managers. Using these new techniques, knowledge is gained in the lab as well as in the field. Physiological data that once could only be collected from animals after their death, can now be monitored while they are alive. Interpretations of behaviors, once largely a subjective and arbitrary exercise, can now be supplemented with visual and audio recordings – including recordings of sights and sounds beyond the range of our own eyes and ears.

New technologies allow researchers better opportunities to study hippos more closely.

Studies, in general, of any species rarely conclude that an animal is less interesting than previously thought. To the contrary, every new study seems to create twice as many questions. With this in mind, and considering how little research has been focused on hippos, the new findings indicate we have probably just scratched the surface of all that there is to learn.

Though recent discoveries may be leading the academic world to a minor epiphany with respect to the importance of hippos and their conservation, this message is still frequently lost on indigenous peoples who must coexist with them. Throughout the African continent, hippos are widely reported to be the number one killer of people. Such a reputation makes grass roots conservation measures difficult (to say the least) to promote. Nevertheless, it is an obstacle that must be overcome through education.

As I have traveled through Africa to discuss and learn about hippos, the two adjectives I most frequently hear to describe the common hippopotamus are 'curious' and 'clever.' Perhaps these descriptions are unexpected to those looking purely for aesthetics. However, it serves as a reminder that hippos should be taken at more than face value. Hippos are beautiful in ways that scientific observations cannot tabulate but researchers are finally beginning to characterize. Hippos offer us insight into their world in ways that scientific observations must sometimes supplement with subjective interpretations. However, the difficulty in measuring an incalculable value should not be a deterrent to researching it further. The effort of learning about and appreciating the things that we do not yet, and possibly will not ever, fully understand, is a contribution to our planet. And, an ever-growing understanding and appreciation of hippos, their purpose, and their history, is a small but very overdue ingredient.

Our current knowledge of hippos represents only the tip of the iceberg.
Studies continue to ensure the future survival of the hippo.

Hippo Populations in Africa

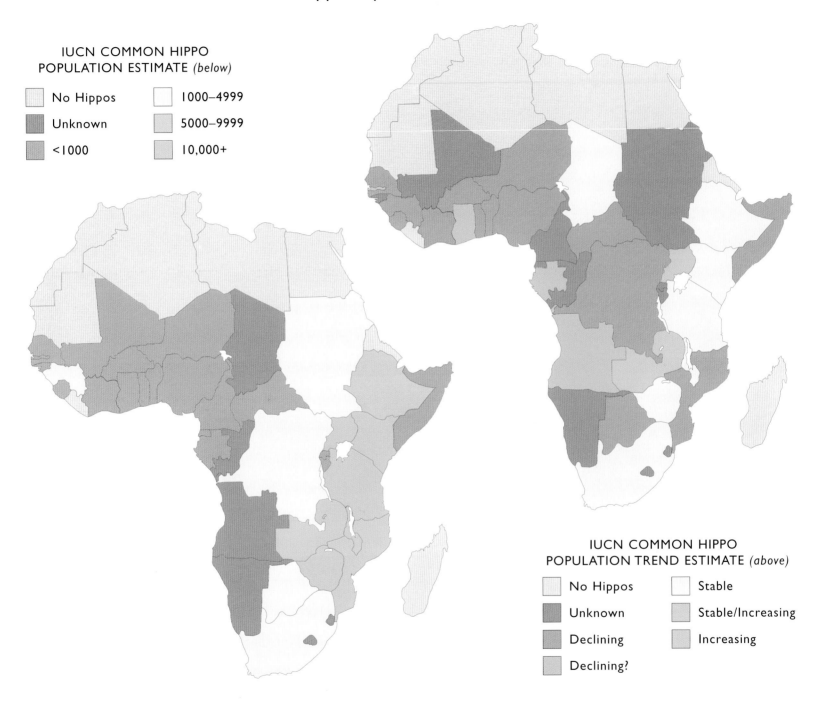

IUCN COMMON HIPPO
POPULATION ESTIMATE (below)

- No Hippos
- Unknown
- <1000
- 1000–4999
- 5000–9999
- 10,000+

IUCN COMMON HIPPO
POPULATION TREND ESTIMATE (above)

- No Hippos
- Unknown
- Declining
- Declining?
- Stable
- Stable/Increasing
- Increasing

Hippo Facts

Common names: Common hippo, Nile hippo, river hippo

Scientific name: *Hippopotamus amphibius*

Similar species: Pygmy hippopotamus (*Hexaprotodon liberiensis*)

Gestation: 8 months

Age at maturity: Males 7-8 years
Females 9-10 years (Both sexes vary with habitat – usually earlier in zoological parks.)

Interbirth interval: 2-3 years

Longevity: 30-40 years (approximately 10 years longer in zoological parks)

Adult weight: Males 1400-2700 kg
(3000-6000 lb)
Females 900-2300 kg
(2000-5000 lb)

Color: Grey, reddish-brown, or nearly black often with lighter pigmentation around eyes and ears.

Distinguishing characteristics: Only smaller than elephants and the largest white rhinos. The lower canine teeth may extend 12 in (30 cm) above the gum line. Hippos have nostrils and ears which can close off when they submerge.

Estimated population (2004): 125,000-148,000 living across at least 29 countries.

Conservation concerns: With the international ban on the trade of elephant ivory, the enormous canines, and occasionally the incisors, of hippos have been used as a legally available and chemically equivalent substitute. Also, habitat loss.

Common names for Hippopotamus
English: Hippopotamus or Hippo   Maasai: Olmakau
Afrikaans: Seekoei   Nyanja: Mvuu   Bambara: Mali
Setswana: Kubuga   Bemba: Mfubu   Swahili: Kiboko

Recommended Reading

Eltringham, S.K. (1999) *The Hippos: Natural History and Conservation*; Princeton University Press.

Estes, R.D., Otte, D., Wilson, E.O.; (1992) *The Behavior Guide to African Mammals: Including Hoofed Mammals, Carnivores, Primates*; University of California Press.

Kingdon, J. (1988) *East African Mammals: An Atlas of Evolution in Africa*, Volume 3, Part B : Large Mammals (East African Mammals); University of Chicago Press.

Oliver, W.L.R. (1993) *Status Survey and Conservation Action Plan: Pigs, Peccaries, and Hippos*; IUCN.

Index

Biographical Note

Glenn Feldhake began his work with common and pygmy hippos as a keeper's aide for seven years at Smithsonian's National Zoological Park in Washington, D.C.. He develops public education materials on hippos and hippo conservation for the Hippo Specialists Group of the World Conservation Union (IUCN). Glenn travels regularly to Africa to work with researchers as well as safari operators. He has distributed the information he has compiled about hippos, their conservation, and environmental influences to researchers, school children, and political leaders alike. His work has also been published in magazines as well as publicized on the radio.

Glenn currently lives in Medina Twp, OH, in the United States where he works as an engineer for the National Aeronautics and Space Administration (NASA).